CASE BOOK

for Moore, McCabe, Duckworth, and Sclove's
The Practice of Business Statistics

WILLIAM I. NOTZ
DENNIS K. PEARL
ELIZABETH A. STASNY

W. H. Freeman and Company
New York

ISBN: 0-71675747-8 (EAN: 9780716757474)

© 2003 by W. H. Freeman and Company

All rights reserved.

Printed in the United States of America.

Second printing

W. H. Freeman and Company
41 Madison Avenue
New York, NY 10010
Houndmills, Basingstoke
RG21 6XS, England

www.whfreeman.com

Contents

	Preface	v
Chapter 1	Examining Distributions	1
Chapter 2	Examining Relationships	6
Chapter 3	Producing Data	11
Chapter 4	Probability and Sampling Distributions	18
Chapter 5	Probability Theory	22
Chapter 6	Introduction to Inference	27
Chapter 7	Inference for Distributions	33
Chapter 8	Inference for Proportions	39
Chapter 9	Inference for Two-Way Tables	46
Chapter 10	Inference for Regression	50
Chapter 11	Multiple Regression	56
Chapter 12	Statistics for Quality: Control and Capability	63
Chapter 13	Time Series Forecasting	67
Chapter 14	One-Way Analysis of Variance	74
Chapter 15	Two-Way Analysis of Variance	80*
Chapter 16	Nonparametric Tests	81
Chapter 17	Logistic Regression	87*
Chapter 18	Bootstrap Methods or Permutation Tests	88

*Case studies are not included for Chapters 15 and 17.

Preface

It is our belief the statistics is best learned by doing. Simply watching someone else solve problems is not an effective way to master material, in much the same way that watching someone else exercise does little to improve one's own fitness. Thus, we have written this case study book so that it can be used as a workbook. We believe that it will be most effective if students actually work through the cases themselves, so we provide space after each question for written answers. The answers can be turned in to an instructor, or they can provide an outline for a comprehensive report.

The case studies in this workbook give students an opportunity to apply some of the statistical methods they have learned in *The Practice of Business Statistics* to solve real problems. There is one case study for each chapter except for Chapters 15 and 17. The cases are based on real examples with real data. In most cases, students will need to use statistical software to carry out their analyses. Not all the statistical methods in a chapter are needed to analyze the data, and some cases can be analyzed in more than one way. In fact, alternate analyses are often possible using methods discussed in other chapters. We have indicated some instances where this is so by follow-up questions at the end of the case.

The cases in this book come from stories in the *Electronic Encyclopedia of Statistical Examples and Exercises* (EESEE). We indicate the story in EESEE from which each case comes. The data for the case are available in electronic form in EESEE. EESEE is available as a web-based supplement to *The Practice of Business Statistics*. It consists of approximately 100 stories based on real examples drawn from a wide variety of disciplines (the humanities, social sciences, biological sciences, medical sciences, physical sciences, business, and sports) that illustrate a wide variety of statistical techniques. The stories come from research journals as well as from the popular media. Many include real data. The cases in this workbook represent only a small fraction of those in EESEE. Students or instructors looking for additional cases or examples will find EESEE a valuable resource. The data associated with stories in EESEE are also a valuable resource and can be used for purposes other than those suggested by the stories. We continue to update EESEE with stories of current interest.

CHAPTER 1

EXAMINING DISTRIBUTIONS

CASE STUDY: Summer Rally

A belief held by many investors is that during the summer the prices of stocks rise. This rise is called the summer rally. It is difficult to find a precise definition of this phenomenon. Does the summer rally apply to all stocks? Just the Dow Jones Industrial Average? One definition of the

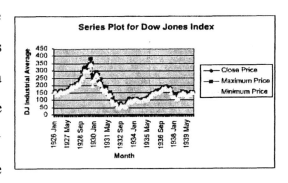

summer rally that has been proposed is the following. First find the lowest closing value of the Dow Jones Industrial Average for the period May through June, the time prior to the start of summer. Next find the highest closing value for the three-month period July through September, the summer months. The difference between these two values is the summer rally.

summer rally =   highest closing value of the Dow Jones Industrial Average over the
                  three month period of July through September

              − lowest closing value of the Dow Jones Industrial Average over the
                  two month period of May and June

To illustrate how the summer rally is computed, on the next page we give the closing price, the maximum price, and the minimum price for the Dow Jones Industrial Average for each month in 1998 and 1999.

We see that the lowest closing value in the period May 1999 through June 1999 is 10466.9 and the highest closing value in the period July 1999 through August 1999 is 11326.0. Thus, the 1999 summer rally is

$$\text{summer rally} = 11326.0 - 10466.9 = 859.1$$

Prevailing wisdom says that this value is generally expected to be positive.

Year	Month	Closing Price	Maximum Price	Minimum Price
1998	Jan	7906.5	7979.0	7580.4
1998	Feb	8545.7	8545.7	8107.8
1998	Mar	8799.8	8906.4	8444.3
1998	Apr	9063.4	9184.9	8868.3
1998	May	8900.0	9211.8	8900.0
1998	Jun	8952.0	9069.6	8627.9
1998	Jul	8883.3	9338.0	8883.3
1998	Aug	7539.1	8786.7	7539.1
1998	Sep	7842.6	8154.4	7615.5
1998	Oct	8592.1	8592.1	7632.5
1998	Nov	9116.5	9374.3	8706.1
1998	Dec	9181.4	9321.0	8695.6
1999	Jan	9358.8	9643.3	9120.7
1999	Feb	9306.6	9552.7	9133.0
1999	Mar	9786.2	10006.8	9275.9
1999	Apr	10789.0	10878.4	9832.5
1999	May	10559.7	11107.2	10466.9
1999	Jun	10970.8	10970.8	10490.5
1999	Jul	10655.2	11209.8	10655.2
1999	Aug	10829.3	11326.0	10646.0
1999	Sep	10337.0	11079.4	10213.5
1999	Oct	10729.9	10729.9	10019.7
1999	Nov	10877.8	11089.5	10581.8
1999	Dec	11497.1	11497.1	10998.4

The EESEE story "The Summer Rally - The Bear Facts" contains an electronic copy of these data, along with additional data sets and descriptions of the variables. You will need the electronic copy to answer the following questions.

QUESTIONS

1. To verify that you understand how the summer rally is computed, use the data in the table to compute the summer rally for 1998.

2. Data set 1 in the EESEE story "The Summer Rally - The Bear Facts" gives the value of the summer rally for each of the years 1960 to 2001. Make an appropriate plot of the distribution of the values. How would you describe the distribution? What is its shape? Are there any outliers? How frequently was the summer rally positive? Is the frequency consistent with prevailing wisdom?

3. Give a measure of the center of the distribution of the values of the summer rally. Is this a reasonable estimate of what we might expect a typical summer rally to be in the next few years? (Hint: Look at how the values of the summer rally have changed over the last few years. Are these values consistent over time or are they changing?)

4   Chapter 1

4. The size of the summer rally may be proportional to the size of the Dow Jones Industrial Average. For example, a summer rally of size 100 in 1960 is, proportionally, a much larger rally than one of size 100 in 1998. To adjust for this proportion, we might consider looking at the percent summer rally, which could be defined as follows:

$$\text{percent summer rally} = \frac{\text{summer rally}}{\text{lowest closing value of the Dow Jones Industrial Average in May or June}} \times 100\%$$

The values of the percent summer rally for 1960 to 2001 are given in data set 2 in the EESEE story "The Summer Rally – The Bear Facts." Make an appropriate plot of the distribution of the values. How would you describe the distribution? What is its shape? What are the center and spread of the distribution? Are there any outliers?

5. Although there is no folklore regarding a winter rally, one could define a winter rally as follows:

winter rally =   highest closing value of the Dow Jones Industrial Average over the three month period January through March
− lowest closing value of the Dow Jones Industrial Average over the two month period November through December

$$\text{percent winter rally} = \frac{\text{winter rally}}{\text{lowest closing value of the Dow Jones Industrial Average in November or December}} \times 100\%$$

The values of the winter rally for the years 1960 to 2001 are given in data set 1, and the values of the percent winter rally for the years 1960 to 2001 are given in data set 2. Make an appropriate plot of the distribution of the percent winter rally values. How would you describe the distribution? What is its shape? What are the center and spread of the distribution? Are there any outliers? How does this distribution compare to that of the percent summer rally? Is the summer rally or the winter rally generally larger?

6. The summer rally and the winter rally were always positive for the years 1960 to 2001. Does this suggest that a wise investment strategy is to invest in the stock market in the late spring, remove the money at the end of the summer, reinvest in the late autumn and again remove the money at the end of the winter? Why or why not? Think carefully about the definition of the summer rally. Under what conditions would there be no summer rally (in other words, when would the formula for the summer rally produce a negative number)? Is a negative number likely to happen?

For additional information and problems related to this case study, see the EESEE story "The Summer Rally - The Bear Facts."

CHAPTER 2

EXAMINING RELATIONSHIPS

CASE STUDY: Faculty Salaries

Universities often compare themselves to other, similar universities to determine if they are competitive on salaries, benefits, tuition, and so on. The 1992 – 1993 Faculty Compensation and Benefits Committee at the Ohio State University (OSU) obtained comparative data on faculty salaries from the Association of American Universities. The average salaries for full professors, associate professors, and assistant professors for each of the top 50 schools in this association were recorded. The goal of the study was to determine if faculty salaries at OSU were comparable to salaries at similar institutions. If salaries at a university fall too low in comparison to those at other schools, many of the good faculty members will move to another school where the salaries are better.

The data are given in the table. For each of the 50 institutions, the average salary (in thousands of 1992 dollars) for full, associate, and assistant professors is provided. In addition, the CIC variable indicates if a school is a member of the Committee on Institutional Cooperation. The 12 member institutions of the CIC include the Big Ten (plus Penn State) and the University of Chicago.

University	CIC	Full	Assoc.	Asst.
DUKE	0	83.00	57.50	46.10
VANDERBILT	0	78.90	49.70	42.50
WASHINGTON UNIV.	0	75.40	51.40	43.80
TULANE	0	70.20	50.80	41.50
CAL TECH	0	93.30	70.00	56.40
CARNEGIE MELLON	0	79.30	55.00	49.40
CORNELL	0	70.96	52.35	44.86
VIRGINIA	0	71.60	47.80	39.50
TEXAS	0	70.30	45.70	40.80
ROCHESTER	0	75.50	51.70	43.90
NEBRASKA	0	63.50	46.00	40.30
UNIV. OF IOWA	1	69.50	50.40	42.80

University				
STANFORD	0	91.20	64.40	50.00
COLORADO	0	64.90	49.60	42.40
UNIV. OF PENN.	0	90.50	64.10	56.20
MICHIGAN	1	73.10	54.00	44.80
PRINCETON	0	92.70	54.90	43.00
IOWA STATE	0	67.10	49.70	40.00
PURDUE	1	70.20	47.90	40.70
UNIV. OF CHICAGO	1	86.90	57.30	50.90
YALE	0	90.20	52.30	43.20
WISCONSIN	1	65.50	48.40	42.60
PENN STATE	1	68.80	49.10	40.30
CAL BERKELEY	0	79.80	53.70	44.80
ILLINOIS	1	67.10	48.40	41.70
MINNESOTA	1	66.50	47.60	41.80
PITTSBURGH	0	71.40	50.80	40.80
HARVARD	0	96.50	55.70	50.00
NORTHWESTERN	1	82.60	57.30	47.60
MISSOURI	0	56.70	42.70	38.70
INDIANA	1	65.70	46.90	38.20
JOHNS HOPKINS	0	76.80	51.40	41.90
CASE WESTERN	0	72.90	50.90	44.30
BROWN	0	72.40	49.40	41.60
M.I.T.	0	87.00	61.60	51.10
MARYLAND	0	73.00	53.70	41.60
OHIO STATE	1	68.80	48.70	41.00
NORTH CAROLINA	0	67.10	48.30	39.50
SYRACUSE	0	63.00	46.10	38.80
MICHIGAN STATE	1	62.40	47.30	38.80
U.S.C.	0	77.90	55.20	46.10
KANSAS	0	55.90	41.80	35.90
U.C.L.A.	0	76.90	51.50	42.90
UNIV. OF WASH	0	66.70	46.90	41.30
OREGON	0	56.10	42.00	35.30
CAL SAN DIEGO	0	75.00	50.50	43.00
CLARK	0	65.30	46.30	40.00
CATHOLIC UNIV.	0	61.00	44.00	37.20
N.Y.U.	0	85.10	56.80	51.00
COLUMBIA	0	83.40	55.80	43.00

The EESEE story "Faculty Salary Comparison" contains an electronic copy of these data, along with additional data sets and descriptions of the variables. You will need the electronic copy to answer the following questions.

QUESTIONS

1. Create scatterplots of salary for full versus associate professors, full versus assistant professors, and associate versus assistant professors. Describe the relationships you see in the plots.

2. Do there appear to be any outliers in any of the plots you created in Question 1? Identify the institutions corresponding to any outliers. Can you explain why these schools might be different from the others?

3. Recreate the scatterplots of Question 1 identifying the CIC schools on the plots. Do you see any differences between the CIC institutions and the other institutions in these plots? Explain

4. Look at the three plots from Question 1. Which plot exhibits the strongest linear association between salaries: full versus associate professors, full versus assistant professors, or associate versus assistant professors? What do you think the correlations are for the relationships in each plot? Obtain the correlations. Are they about what you expected?

5. What do you think would happen to the correlation between salaries for full and associate professors if you deleted Harvard, Princeton, and Yale from the data? Obtain this correlation. Did the correlation change in the way you expected? Explain.

6. Obtain the least squares regression line for predicting full professors' average salary using assistant professors' average salary. What does the slope of this regression line tell you? What does the intercept of this regression line tell you?

7. About how much variation in full professors' average salaries is explained by the regression on assistant professors' average salaries? How is this quantity related to the correlation between full and associate professors' average salaries that you found in Question 4?

8. Consider a school that had an average assistant professors' salary of $50,000 in 1992 – 1993. What does the regression line predict as the average salary for full professors? Look at the schools in the data that have average salaries of $50,000 for their assistant professors. What are their actual average salaries for full professors? How do these values compare to the predicted value?

9. Plot the residuals from the least squares regression line versus average assistant professors' salaries. Do you see any evidence of outliers? Of a curved pattern? Is there any evidence of influential observations?

For additional information and problems related to this case study, see the EESEE story "Faculty Salary Comparison."

CHAPTER 3

PRODUCING DATA

CASE STUDY: Does Use of an Incentive Increase Response Rates for Mailed Surveys?

Mailed surveys are often used to collect information about issues such as consumer satisfaction, use of community services, and public opinions. A disadvantage of mailed surveys is that response rates can be quite low. Potential respondents may view the survey as junk mail and simply throw it out. Therefore, survey research organizations seek methods that will increase the number of respondents who fill out the survey and mail it back. One possible method to increase response rates is to offer respondents a monetary incentive for completing a survey. Three studies conducted in 1992 in New Zealand found that including an incentive with the mailed survey could increase the response rate.

One survey asked about the potential use of a new sports facility. The second survey asked dairy and beef farmers about their use of mineral supplements. A third survey of individuals sampled from lists of registered voters asked about social inequalities. The researchers sent up to three mailings in each of the three studies. In the first mailing, the survey went to all individuals in the sample and included the survey and the incentive (if there was one). In the second and third mailings the survey was resent again to people in the sample who had not already returned their surveys.

The following three tables show the total number of respondents after each mailing of each of the three surveys using various types and amounts of incentives. Notice that the numbers of respondents represent the *cumulative* numbers of respondents at the end of each wave.

Chapter 3

Sports Survey

This survey looked at potential usage of a proposed sports facility. The researchers randomly assigned 350 people to one of four conditions: control (no incentive), an incentive of 20 New Zealand cents, an incentive of 50 cents, or an incentive of one dollar. N is the number of surveys in each condition that were not returned because no one lived at the address any longer.

Cumulative Numbers of Respondents to Each Mailing for the Sports Survey

	Control	20 New Zealand cents	50 New Zealand cents	1 New Zealand dollar
N	79	79	80	82
Mailing 1	21	23	32	32
Mailing 2	35	42	46	49
Mailing 3	44	49	53	56

Farmers Survey

Individuals from a national sample of 250 dairy and 250 beef farmers were randomly assigned to one of four conditions: control (no incentive), an incentive of an Instant Kiwi lottery ticket, an incentive of 50 cents, or an incentive of one dollar. This assignment was done separately for the dairy and beef farmers. The survey asked about the use of mineral supplements for the animals. N is the number of surveys in each condition that were not returned because no one lived at the address any longer.

Cumulative Numbers of Respondents to Each Mailing for the Farmers Survey

	Control	Instant Kiwi (equal to 1 New Zealand dollar)	50 New Zealand cents	1 New Zealand dollar
N	101	108	103	100
Mailing 1	27	35	42	35
Mailing 2	38	55	59	51
Mailing 3	50	64	70	66

Voters Survey

A random sample of 2154 individuals selected from the electoral roll were randomly assigned to one of four conditions: control (no incentive), an incentive of 50 cents, an incentive of one dollar, or an incentive of a one-dollar donation to a charity. The survey looked at attitudes about social inequality. N is the number of surveys in each condition that were not returned because no one lived at the address any longer.

Cumulative Numbers of Respondents to Each Mailing for the Voters Survey

	Control	50 New Zealand cents	1 New Zealand dollar	Donation of 1 New Zealand Dollar
N	452	454	463	478
Mailing 1	127	183	208	168
Mailing 2	207	255	270	239
Mailing 3	230	278	307	271

Chapter 3

Note: Following is the approximate conversion rate from New Zealand to U.S. money based on 1992 conversion rates:
- 20 New Zealand cents = 11 U.S. cents
- 50 New Zealand cents = 27 U.S. cents
- 1 New Zealand dollar = 54 U.S. cents

QUESTIONS

1. What is the population of interest for each of the three surveys?

2. For each of the three surveys, describe possible biases that could result from a voluntary response sample.

3. Which of the three surveys uses a stratified sample? What are the strata?

4. Why did the researchers send out three mailings of the surveys? Why did the researchers contact the same respondents in mailings 2 and 3 instead of sending the surveys to a new sample of individuals?

5. For the sports survey, determine the percentage of nonresponse for each of the four conditions. Do you think nonresponse might be a problem in this survey? Why?

6. Although the main data collection method for the three studies is a sample survey, there are actually experiments embedded in each survey. What are the treatments for each of the three studies?

7. Do these studies follow the principles for a good experimental design? Explain why or why not.

8. For each study, compute the percentages of individuals who ultimately responded to the survey under each incentive condition. Does the size or type of incentive seem to make a difference in the response rate?

9. In addition to the goal of increasing survey response rates, one might consider using incentives to reduce costs of the survey. Consider the conditions in the farmers survey. Assume that it costs a dollar to mail the survey and you are mailing the survey to 100 people. Suppose you get the following responses under the various treatment conditions: for the control 27 surveys are returned in the first mailing and another 11 in the second mailing; for the Instant Kiwi incentive 32 are returned after the first mailing; for the 50 cents incentive 41 are returned after the first mailing, and for the one-dollar incentive 35 are returned after the first mailing. Would it be less expensive to send out a second mailing or to include an incentive?

10. Do you think the results concerning incentives from these studies can be applied to any mailed survey? Explain why or why not.

For additional information and problems related to this case study, see the EESEE story "Mail Survey Incentives."

CHAPTER 4
PROBABILITY AND SAMPLING DISTRIBUTIONS

CASE STUDY: Baby Hearing Screening

According to the National Campaign for Hearing Health, deafness is the most common birth defect. Audiologists estimate that three out of every thousand babies are born with some kind of hearing loss. Early identification of hearing loss through a screening test shortly after birth and subsequent intervention can have a significant impact on a child's cognitive development as well as having financial consequences. Thus, the Universal Newborn Hearing Screening bill calls for high-performing, accurate, and inexpensive testing equipment. The screening tests work by introducing a sound into a baby's ear and then measuring either the response of the ear's internal mechanisms or the electrical activity of the auditory portion of the brain. Just because a baby fails those tests, however, does not mean that there is a hearing problem.

Studies conducted at the University of Utah were undertaken to assess the adequacy of a new miniature screening device called the Handtronix-OtoScreener (a small hand-held audiometer developed by the Utah Innovation Center, Salt Lake City, UT). The accuracy of this new device was judged using as "truth" the results found using standard equipment. Results are shown in the following table:

	Baby's Hearing ("Truth")	
	Loss	Normal
Test Result	54	6
	4	36

QUESTIONS

1. Sensitivity and specificity are measures of the accuracy of a diagnostic procedure. For the Handtronix-OtoScreener, sensitivity is defined as the probability that the test shows a hearing loss when there really is a hearing loss. Specificity is defined as the probability that the test shows that hearing is normal when hearing is really normal.

(a) Estimate the "sensitivity" and the "specificity" of the new small audiometer using the data in the table.

Estimate of sensitivity = proportion of those babies with a hearing loss who show a hearing loss when tested

=

Estimate of specificity = proportion of those babies with normal hearing who test normal

=

(b) Estimate the chance that a baby with a hearing loss will pass the hearing test (test normal) using the new miniature screening device.

Estimate = proportion of those babies with a hearing loss who test normal

=

2. It is estimated that the total number of births in the U.S. each year is about 4,008,083 and the prevalence of hearing loss is three per thousand. Based on these figures, how many babies each year can be expected to have a hearing loss?

3. Suppose all babies born in the U.S. this year are screened for a hearing loss. Assume that this number of births is 4,008,083, as in Question 2. Using the information in the table, estimate the number of babies with a hearing loss that will be missed by the new screening device.

4. Assume that the probability of a hearing loss in infants is three per thousand, i.e., 0.003. Suppose we choose a baby randomly among all newborns and test the baby's hearing using the Handtronix-OtoScreener. If the test shows that the baby has a hearing loss, what would you estimate to be the probability that the baby really has a hearing loss?

5. Now let's look at the cost of a newborn hearing screening program. For simplicity, assume that 2000 babies are born per year in a large hospital and that 100% of those babies are screened. Again, assume that the probability of a hearing loss is 0.003. The annual cost of a screening program for 2000 babies using one type of hearing screening equipment (Natus Algo equipment) is $46,735. What should the family of each baby be charged to cover this cost? Suppose one were to charge only families whose babies are identified as having a hearing loss. How many of the 2000 babies would you expect to be identified as having a hearing loss and what would you have to charge the families of these babies to cover the $46,735 cost?

For additional information and problems related to this case study, see the EESEE story "Baby Hearing Screening - The Sound of Silence."

CHAPTER 5
PROBABILITY THEORY

CASE STUDY: How Likely Is Your Flight to Be On Time?

Airlines are required to make monthly reports on the percentage of flights that arrive on time at the nation's 30 busiest airports. These figures are often cited by airlines in their advertisements as a testament to their performance. In one example, a June 1991 report indicated that America West Airlines had a higher percentage of on-time-flights, 89.1%, than did Alaska Airlines, 86.7%, for the airports served by both airlines. We can use probability theory to help us understand what these percentages mean to travelers. We can also use statistical thinking to help us assess the truthfulness of such advertising.

The table below provides the June 1991 on-time performance comparison for the flights arriving at airports in cities serviced by both America West Airlines and Alaska Airlines.

Destination	Alaska Airlines Number On Time	Total Flights to City	America West Airlines Number On Time	Total Flights to City
Los Angeles	497	559	694	811
Phoenix	221	233	4840	5255
San Diego	212	232	383	448
San Francisco	503	605	320	449
Seattle	1841	2146	201	262
All five cities	3274	3775	6438	7225

QUESTIONS

1. Verify that the proportions of on-time flights for Alaska Airlines and America West Airlines in June 1991 are as quoted in the advertisement.

2. Compute the proportions of on-time flights for each airline for each of the five destinations shown in the table. Suggest reasons for the differences you see in the proportions for each city.

3. Suppose you work for a company that requires you to travel to the five cities serviced by Alaska Airlines and America West Airlines. Use the proportions from Question 1 as if they were the population proportions. If you fly Alaska Airlines on your next four flights, what is the probability that all four flights will be on time? What is the probability if you fly America West Airlines?

4. What is the probability that at least one of your next four flights will be late if you fly Alaska Airlines? If you fly America West Airlines?

5. What is the probability that exactly two of your next four flights will be on time if you fly Alaska Airlines? If you fly America West Airlines?

6. Suppose your company requires you to travel to San Francisco by either Alaska Airlines or America West Airlines. If you fly Alaska Airlines on your next four flights, what is the probability that at least one of those flights will be late? What is the probability if you fly America West Airlines?

7. Over the next year, employees of your company will make 100 trips to San Francisco on either Alaska Airlines or America West Airlines. What is the mean number of on-time flights out of 100 flights on Alaska Airlines? What is the standard deviation? What are the mean and standard deviation for 100 flights on America West Airlines?

8. What is the approximate probability that at least 90 of 100 flights to San Francisco on Alaska Airlines will be on time? What is the approximate probability that at least 85 of 100 such flights will be on time?

9. The probabilities that you calculated in Questions 3 through 8 were based on data from June 1991. What assumptions must you make for the probability calculations to be valid? Do you believe the assumptions?

10. Using the June 1991 data, compute the proportion of flights that Alaska Airlines flew to Seattle and the proportion of flights that America West Airlines flew to Seattle. Consider the differences between Seattle and the other four cities. Are there any implications concerning the truthfulness of the comparison of the proportions of on-time flights cited in the advertisement?

For additional information and problems related to this case study, see the EESEE story "On-Time Flights."

CHAPTER 6
INTRODUCTION TO INFERENCE

CASE STUDY: Precision Time

Time keeping made a great leap forward in the 17th century when Christiaan Huygens built the first pendulum clock. The idea was simple — the regular back-and-forth motion of the pendulum was so repeatable, you could tell time by just counting the swings. An accurate method of keeping time was crucial to the advancement of worldwide trade because navigation depended on knowing the position of the sun at a specific time. Unfortunately, the first pendulum clocks did not work aboard ships, so in 1714, the British government offered a £20,000 prize (about $2,000,000 today) for a timing device that would cause an error of less than a half degree of longitude on a trip to the West Indies. The prize was won by John Harrison in 1761 with a clock that could withstand the movement of a ship and still produce an error of less than a half second per day. That was ten times better than required.

Today, the world's most accurate clock is the NIST-F1 built by the National Institute of Standards and Technology in Boulder, CO. It is an atomic clock that keeps time by counting the transitions of the cesium atom as it moves back and forth between two energy levels. It is accurate to a tenth of a nanosecond: it might lose a second in 20 million years! An important purpose of the NIST-F1 is the same as the purpose of John Harrison's maritime clock — commerce and navigation. Calculating bank transfers, conducting space travel, synchronizing television feeds, and transmitting e-mail all require precision time. The NIST-F1 is also used to calibrate the atomic clocks aboard the satellites used in global positioning systems (GPS) that could otherwise be off by a dozen nanoseconds. A nanosecond (one billionth of a second) seems small, but a nanosecond of error in an advanced GPS translates into a navigational error of about a foot. Pilots using GPS instruments to land on an aircraft carrier are thankful for every nanosecond of accuracy!

The NIST-F1 makes contact on a second-by-second basis with each of the satellites in the Global Positioning Network that it can detect in the sky. The differences between a satellite's onboard clock and the NIST-F1 are then reported as averages over ten-minute periods. Since each satellite can communicate with Boulder, CO, for about 40 ten-minute periods per day, there are around 30,000 of these values posted each month for the network as a whole. A random sample of 100 of the ten-minute averages for the month of December 2002 is given in the table on the following page.

The sample in the table excludes December 19 and 20 when NIST did not post the values. Day of the month is given with times on the universal 24-hour clock (for example, "27,20:50" indicates a ten-minute average ending at 8:50 p.m. in Greenwich, England, on December 27, 2002). Negative error values indicate that the satellite clock was running slower than NIST-F1 while positive values indicate that it was running faster.

Day, Time	Satellite ID	Error (ns)	Day, Time	Satellite ID	Error (ns)	Day, Time	Satellite ID	Error (ns)
21,07:10	3	6.95	13,02:00	10	-12.75	18,22:50	28	-0.40
30,11:20	20	16.20	6,02:00	10	6.90	28,21:50	24	-20.15
4,19:40	28	-5.10	24,13:30	13	10.90	2,15:30	22	0.25
16,03:00	5	-2.90	29,23:50	4	9.65	23,14:30	25	-6.5
6,05:10	17	7.95	29,16:40	13	6.25	22,04:10	18	8.35
14,09:20	18	-9.5	17,09:20	31	4.00	16,11:00	25	3.15
10,19:40	8	-19.9	26,13:50	1	10.35	3,22:50	9	11.95
27,14:30	20	2.35	5,03:00	24	4.50	15,02:00	30	-20.55
1,20:50	26	5.25	15,15:30	1	-24.8	26,14:50	27	_0.20
30,06:10	17	14.00	2,10:20	9	6.00	30,23:50	24	1.40
25,22:50	28	-4.70	17,00:00	4	-4.45	15,08:20	17	-7.60
5,19:40	27	-0.05	3,18:40	27	4.00	22,04:10	29	11.35
25,23:50	4	-1.50	25,11:20	30	7.00	28,15:30	27	6.20
15,21:50	7	-0.40	9,10:20	9	-6.10	25,05:10	26	8.55
26,09:20	25	5.05	28,08:20	31	11.35	23,01:00	30	-5.70
4,02:00	4	7.85	27,12:30	11	1.35	1,18:40	2	-5.40
5,07:10	15	3.80	31,01:00	24	4.25	6,03:00	4	15.65
11,10:20	31	-13.40	18,19:40	8	-9.75	17,19:40	7	-3.10
2,22:50	26	9.35	1,03:00	4	12.95	23,23:50	4	6.75
31,15:30	27	11.15	24,05:10	6	-3.30	6,20:50	26	13.45
31,10:20	25	4.10	18,20:50	29	-0.80	28,15:30	13	5.05
14,03:00	30	-14.15	26,16:40	20	-7.65	16,07:10	18	-1.00
27,01:00	24	-2.5	17,15:30	31	-40.9	1,18:40	8	1.25
8,16:40	27	-18.95	9,07:10	6	-5.00	26,08:20	14	3.40
1,14:30	11	1.90	12,00:00	9	-6.15	23,13:30	1	17.95
26,10:20	30	7.50	22,07:10	26	15.35	6,19:40	7	-9.55
27,20:50	4	-17.75	8,00:00	20	-3.80	8,21:50	11	-9.45
11,15:30	13	-19.95	28,23:50	24	-6.85	21,13:30	20	6.15
24,13:30	1	23.55	18,04:10	6	-4.10	9,17:40	27	-22.65
12,23:50	7	-8.60	11,01:00	24	8.6	23,21:50	28	0.85
11,16:40	27	-17.65	31,09:20	14	14.40	22,08:20	14	9.2
14,22:50	28	-26.5	9,05:10	15	-8.95	1,09:20	31	12.30
27,01:00	24	-2.50	31,05:10	26	17.90	6,15:30	20	-11.65
18,14:30	1	8.45						

QUESTIONS

1. Coordinated Universal Time (UTC) is the official world time. UTC is a weighted average of 250 different atomic clocks in 50 countries. Why not just use the best clock (NIST-F1) to set the official world time? There may be a political reason — but there's also a statistical reason. Explain.

2. To win the 1714 British navigation prize, you needed to convince the Board of Longitude of the Royal Society that you had produced a clock that made an error of less than 5 seconds per day on trips to the West Indies.

a) Suppose someone was trying to defraud the Royal Society and had a clock that was so variable as to be clearly undeserving of the prize. For example, suppose the error in the time on this clock followed a normal curve with a mean of zero and a standard deviation of 25 seconds per day on the two-month trip to the West Indies. What is the chance that this clock could make the trip and pass the test?

b) At first, the Board of Longitude refused to authorize the payment of the 1714 prize to John Harrison despite the fact that his clock successfully completed the challenge in 1761. In fact he wasn't fully paid until 1773, after several other successful trials and the intervention of King George III. Was the Board of Longitude too skeptical, or was there good reason to hold off on the payment? There may have been a political reason — but there's also a statistical reason. Explain.

3. For December 2002 as a whole, the ten-minute averages from individual satellites have a standard deviation of about 10 nanoseconds. Estimate the average error made by the ten-minute averages from the Global Positioning Satellite Network in December. Calculate a 90% confidence interval for this average and explain what it means. How would your interval change if 95% confidence were required?

4. Did the Global Positioning Satellite Network send an unbiased estimate of the time to GPS systems in December 2002? What would be the appropriate null hypothesis to address this question with a significance test? Would you use a one-sided or a two-sided alternative hypothesis? Explain.

5. Carry out the significance test from Question 4. Be sure to calculate the test statistic and the P-value. How do you interpret this P-value? What conclusions can you make about the satellite network?

6. If every value in the table were 3 nanoseconds lower, there would be clear evidence that the satellite network was biased (prove this to yourself). Would that prove that the Global Positioning Network was too biased in December to be useful to a person driving a car with a GPS trying to find an intended destination? Explain the statistical lesson.

For additional information and problems related to this case study, see the EESEE story "Precision Time."

CHAPTER 7
INFERENCE FOR DISTRIBUTIONS

CASE STUDY: Secondhand Smoking Bartenders

During the past two decades, many studies have been conducted to explore the association between secondhand smoke (sometimes called "environmental tobacco smoke") and lung cancer. Most of these studies concluded that secondhand smoke is strongly related to lung cancer and other lung problems. Bar and tavern workers, in particular, are exposed to unusually high levels of tobacco smoke every day. According to the New York City Department of Health and Mental Hygiene, bartenders working an eight-hour shift inhale cancer-causing chemicals equivalent to smoking a half pack of cigarettes. This finding led the state of California to pass a bill prohibiting smoking in bars and taverns. The law took effect on January 1, 1998.

Smoking bans have both health and economic consequences. In a 1993 report, the World Health Organization estimated that the use of tobacco results in a net loss of $200 billion per year through direct health care costs and loss of productivity. The economic consequences of smoking bans on small businesses like taverns however, are more controversial. Much of the debate focuses on the actual effects of secondhand smoke. Has inhaling secondhand smoke done observable damage to bartenders' lungs? This question was addressed by researchers at the Cardiovascular Research Institute of the University of California, who studied bartenders in San Francisco before and after the California smoking ban took effect.

A random sample of bars and taverns was selected in San Francisco, CA. The sampled bars and taverns that agreed to participate in the study employed 67 bartenders who worked at least one daytime shift per week. The 67 bartenders were the intended subjects in the study. The subjects underwent a standard baseline interview, conducted by a single study investigator, in their workplaces during December 1997. Questions used in this interview asked about respiratory symptoms like coughing and wheezing and about sensory irritations like sore throats.

The follow-up interviews, conducted eight weeks after the law went into effect, asked the same questions about respiratory and sensory irritation problems.

At both interviews subjects were also asked to perform three breathing maneuvers and their lung function was measured with a portable machine called a spirometer. The three lung measurements taken were forced expiratory volume (FEV; amount of air exhaled forcefully from the lungs in one second), forced vital capacity (FVC; amount of air that can be expelled from lungs filled to capacity, with no limit to the duration of expiration), and forced expiratory flow (FEF; rate of airflow leaving the lungs during the period midway through exhaling). An increase in FEV, FVC, or FEF would reflect improved lung function. Fifty-three of the 67 bartenders (79%) completed both interviews and the two lung function tests. Here are the results:

Respiratory Problems (Coughing, Wheezing, Breathing Difficulties, Excess Phlegm)

	Symptoms at Baseline	No Symptoms at Baseline	Total
Symptoms at Follow-up	16	1	17 (32%)
No Symptoms at Follow-up	23	13	36 (68%)
Total	39 (74%)	14 (26%)	53 (100%)

Sensory Problems (Irritation of Eye, Nose, or Throat)

	Symptoms at Baseline	No Symptoms at Baseline	Total
Symptoms at Follow-up	9	1	10 (19%)
No Symptoms at Follow-up	32	11	43 (81%)
Total	41 (77%)	12 (23%)	53 (100%)

Lung Function Tests

Measurement	Baseline $\bar{x} \pm s/\sqrt{n}$	Follow-up $\bar{x} \pm s/\sqrt{n}$	Change $\bar{x} \pm s/\sqrt{n}$
FEV (liters)	3.38 ± 0.13	3.42 ± 0.14	0.039 ± 0.0024
FVC (liters)	4.43 ± 0.15	4.62 ± 0.17	0.189 ± 0.0037
FEF (liters/sec.)	3.37 ± 0.19	3.18 ± 0.17	-0.190 ± 0.0075

QUESTIONS

1. Did the lung function of the bartenders change after the smoking ban? Explain. Describe the population and define the parameter(s) of interest in this situation.

2. Did the amount of air that the bartenders could exhale forcefully from their lungs in one second change significantly after the law went into effect? Explain. Would it be more appropriate to use a one-tailed significance test or a two-tailed test to address this question? Explain.

3. Did the total amount of air that the bartenders could exhale from their lungs change significantly after the law went into effect? Use an appropriate statistical procedure to address this question.

4. Did the rate of air leaving the lungs while exhaling change significantly after the law went into effect? Use an appropriate statistical procedure to address this question. Does it matter whether you use a one-tailed or two-tailed test? Explain.

5. How much did the total amount of air exhaled by the bartenders change from baseline to follow-up? Make a 90% confidence interval to address this question and explain its meaning.

6. The San Francisco bartenders study was published in the December 9, 1998, issue of the *Journal of the American Medical Association*. The authors of the study wrote their conclusion concisely: "Establishment of smoke-free bars and taverns was associated with a rapid improvement of respiratory health." Does the study data support this conclusion? Explain why or why not.

7. Twenty-nine of the bartenders in this study reported no difference in respiratory symptoms over the period of the study, either reporting no symptoms before and after the law went into effect or reporting symptoms at both interviews. Of the remaining 24 bartenders, 23 reported improvement in respiratory symptoms and one reported symptoms after the smoking ban but none beforehand. Explain how these data may be used to evaluate whether the population of San Francisco bartenders felt their respiratory problems were reduced after the smoking ban became law.

For additional information and problems related to this case study, see the EESEE story "Second Hand Smoking Bartenders."

FOLLOW-UP: FACULTY SALARIES (Continued)

Are faculty salaries higher at private colleges than at public colleges? Classify each of the colleges in the faculty salary data set given with the case study for Chapter 2 as "public" or "private" and use an appropriate statistical technique to investigate this question. Are the assumptions underlying your analysis reasonable to make? What do you conclude?

CHAPTER 8
INFERENCE FOR PROPORTIONS

CASE STUDY: Drive-Through Competition

The pork sandwiches at The Pig Stand were a big success as soon as the restaurant opened in Dallas, TX in 1921. As the crowds grew and traffic jams began to form outside, The Pig Stand's waiters would run out to the curb, hop onto the running boards of cars, and take the food orders. This first drive-in restaurant also pioneered the first drive-up window ten years later when The Pig Stand let motorists drive up to the restaurant and place their orders directly. The modern drive-through window started in San Diego, CA in 1948 when the owners of the In-N-Out Burger restaurant installed a two-way speaker so that customers could order from an outdoor menu before pulling up to a window to receive their food. In 1951 fellow San Diego businessman Robert Peterson adapted the idea for his Jack in the Box restaurant, which developed into the first national food chain using the drive-through idea.

Today, over $70 billion is spent each year in the drive-through lanes of America's fast food industry. Having fast, accurate, and friendly service at a drive-through window translates directly into revenue. According to Jack Greenberg, the CEO of McDonald's, sales increase 1% for every 6 seconds saved at the drive-through. Thus, industry executives, stockholders, and analysts closely follow the ratings of fast food drive-thru lanes that appear annually in *QSR* magazine (a trade publication of the quick service restaurant business).

The 2002 *QSR* magazine drive-through study was conducted by g3 Mystery Shopping, a market research firm in Sylvania, OH. The study involved a total of 7594 visits to restaurants in the 25 largest fast food chains in all 50 states. Visits occurred during the lunch hours of 11:00 A.M. to 2:30 P.M. or during the dinner hours of 4:00 P.M. to 7:00 P.M. Four aspects of drive-through operations were measured at each visit: speed, accuracy, menuboard appearance, and speaker clarity. The speed and accuracy measurements are particularly important since together they make up 80% of the final overall rating. Below are the 2002 study data.

Chain name	Number of visits	Service time (seconds)	Inaccuracies (%)	Passing Menuboards (%)	Unclear speakers (%)
A&W	152	204.89	10.53	89.47	18.42
Arby's	290	176.54	11.72	91.72	15.97
Burger King	742	173.37	11.86	94.34	8.94
Captain D's	188	243.93	13.83	93.62	7.45
Carl's Jr.	220	221.42	20.00	89.09	17.27
Checkers	170	173.56	16.47	95.29	13.25
Chick-fil-A	196	150.57	7.14	93.88	3.06
Church's Chicken	208	235.44	18.27	87.50	26.21
Dairy Queen	272	221.75	17.65	91.18	19.85
Del Taco	154	185.32	15.58	93.51	14.29
El Pollo Loco	110	215.63	14.55	96.36	14.55
Hardee's	284	182.69	23.24	95.77	4.93
Jack in the Box	270	223.17	11.85	92.59	7.41
KFC	730	198.72	17.26	89.32	14.29
Krystal	168	170.78	23.81	91.67	7.14
Long John Silver's	240	193.44	16.67	84.17	12.50
McDonald's	750	162.72	12.00	96.00	9.30
Popeyes	248	208.04	14.52	87.90	17.74
Rally's	158	182.67	12.66	98.73	19.23
Steak n Shake	118	269.50	11.86	91.53	8.47
Taco Bell	736	167.15	11.14	92.39	14.17
Taco John's	150	208.53	17.33	90.67	16.22
Wendy's	734	127.21	9.54	96.46	9.02
Whataburger	186	268.15	17.20	94.62	8.79
White Castle	120	240.23	18.33	85.00	15.00
All Chains	7594	186.87	14.12	92.63	12.26

The service time is the average time it took from a customer (the researcher) stopping at the speaker to that customer receiving the entire order. During each visit, the researcher ordered a modified main item (for example, a hamburger with no pickles), a side item, and a drink. If any item was not received as ordered, or if the restaurant failed to give the correct change or supply a straw and a napkin, then the order was considered "inaccurate." A menuboard was considered as "passing" if it had no missing panels or damaged faces, no handwritten signs, and no decals or stickers on its face and if it was generally clean. A speaker was considered unclear if it had an excessive amount of static or a volume level that was too high or too low, was out of order, or was simply unclear.

QUESTIONS

1. The 1998 *QSR* survey showed that McDonald's had an average service time of 177.59 seconds. Since that time many McDonald's restaurants have implemented incentive programs that give employees rewards for fast drive-through times. In 2001, Burger King instituted several measures designed to help employees more accurately fill drive-through orders. For example, they tested see-though bags at some locations. Both these chains would like to know if the programs they instituted had an effect. What are the parameters of interest in each situation? Explain.

2. In the 1998 *QSR* study, Arby's average service time was just over 200 seconds. As a result, Arby's CEO Michael Howe instituted programs to increase speed and vowed to bring the average drive-through service time for all Arby's to below 3 minutes. For example, soft drink machines were moved so that employees could take orders and pour drinks at the same time. At the same time, Arby's did not want to increase the variability from restaurant to restaurant, which is reflected in a service time standard deviation of 17 seconds. Use the data from the 2002 study to test if Arby's has achieved its goal.

3. There are approximately 13,000 McDonald's restaurants in the United States. Suppose you wish to estimate the proportion of orders that would be handled inaccurately if every one of these locations had been visited as part of the 2002 *QSR* study. Give the Wilson estimate of this proportion and a 90% confidence interval.

4. In planning next year's study, suppose g3 Mystery Shopping would like to have less than a 3% margin of error for estimating the inaccuracy rate at McDonald's with 90% confidence. How many visits will be needed?

5. The speaker clarity data in the table are given as the percent of restaurants in the chain that failed the QSR test while the menuboard data give the percent of restaurants that passed. Which method of data presentation is better for the use of the normal approximation in the z-procedures? Or are they the same? Explain.

6. In the 2001 *QSR* drive-through study, 730 of the 890 Burger King restaurants visited gave correct change and accurately filled the order. Was there a significant improvement in accuracy between 2001 and 2002 following Burger King's corporate emphasis on order accuracy? Carry out the appropriate statistical procedure to address this question and write up your results in an executive summary for Burger King management.

7. Combining the four aspects of drive-through quality, g3 Mystery Shopping found Chick-fil-A to have the highest overall ranking in 2002. For example, Chick-fil-A had the lowest inaccuracy rate of any of the chains. Give a 95% confidence interval for the difference between the inaccuracy rate at Chick-fil-A and the rate at McDonald's. Explain the meaning of this interval.

Additional information and problems related to this case study will be available in the EESEE story "Drive-Thru Competition" in the near future.

FOLLOW-UP: HEARING LOSS IN BABIES (Continued)

Use the University of Utah data given with the case study for Chapter 4 to find a 90% confidence interval for the chance that a baby with a hearing loss will pass the hearing test using the new miniature screening device described in the case study.

CHAPTER 9
INFERENCE FOR TWO-WAY TABLES

CASE STUDY: Does Use of a Warming Blanket During Surgery Reduce Cost for Patient Care?

Companies providing health insurance want to ensure that their subscribers receive the most economical treatment possible as well as adequate treatment. One way to control healthcare costs associated with surgery is to reduce the amount of time a patient must spend in the hospital after surgery. Because patients may be required to spend additional time in the hospital if they contract postoperative infections, it could be to an insurance company's advantage to promote hospital procedures that reduce the risk of infection.

When patients undergo surgery, the operating room is kept cool so that the physicians in their heavy gowns will not be overheated. Patients and their insurance companies could pay the price for the surgeons' comfort. The exposure to cold, in addition to the impairment of temperature regulation caused by anesthesia and altered distribution of body heat, may result in mild hypothermia, a reduction of body temperature by approximately two degrees Celsius below the normal core body temperature. As a result of the hypothermia, patients may have an increased susceptibility to wound infections or even morbid cardiac events.

A study conducted in Austria investigated whether using a newly developed warming blanket to maintain a patient's body temperature close to normal decreases wound infection rates. All patients included in the study were undergoing colon or rectal surgeries, which are typically associated with a high risk of infection. Random assignment of patients to either the warming blanket or regular (control) blanket was done during the induction of anesthesia.

The following two-way table provides counts of patients who did and did not suffer postoperative infections by whether they received the warming or regular blanket during surgery.

	Warming blanket	Regular blanket
Infection	6	18
No Infection	98	78

QUESTIONS

1. Find the marginal totals and table total for the above two-way table of counts of patients with and without infections for the treatment and control groups.

2. What percentage of patients who had surgery with the warming blankets suffered an infection? What percentage of patients who had surgery under standard conditions suffered an infection? Does it appear that the warming blanket reduced the rates of infection after surgery?

3. Explain what null hypothesis you would use the X^2 statistic to test in this table.

4. Obtain the expected cell counts, the X^2 statistic, and the P-value for the hypothesis you wish to test.

5. What do you conclude from the test in Question 4? Does it appear that the use of a warming blanket during surgery could reduce postoperative wound infection and, hence, costs associated with longer hospital stays?

6. Compare the observed and expected cell counts for this table. What deviations do the data show from what you would expect under the null hypothesis?

7. One observed cell count is small. Do the data satisfy the guidelines for safe use of the chi-square test?

8. The Austrian study was carried out on patients who were undergoing colon or rectal surgeries, which are typically associated with a high risk of infection. What would this mean for an insurance company considering whether or not to require hospitals to use a warming blanket during surgery on patients they insure?

For additional information and problems related to this case study, see the EESEE story "Surgery in a Blanket."

CHAPTER 10

INFERENCE FOR REGRESSION

CASE STUDY: What's Driving Car Sales?

U.S. car sales change month by month because of car prices, the national economy, and other factors that affect the sales. What about gas price? Does it also affect car sales? To explore this question, we examine U.S. car sales data from January 2000 to June 2001, a period when gas prices were generally increasing. A table containing some relevant data follows.

Month	Total	Sales			Percentage of Sales			Gas Price	
		Low mpg car group	Middle mpg car group	High mpg car group	Low mpg%	Middle mpg%	High mpg%	Gas price current month	Gas price previous month
Jan	406004	155371	155212	95421	38.27	38.23	23.50	130.8	*
Feb	505861	216045	174679	115137	42.71	34.53	22.76	138.9	130.8
Mar	550253	227763	182585	139905	41.39	33.18	25.43	155.0	138.9
Apr	507937	207018	167878	133041	40.76	33.05	26.19	145.5	155.0
May	556589	221505	190979	144105	39.80	34.31	25.89	150.5	145.5
Jun	556137	228046	185452	142639	41.01	33.35	25.65	164.0	150.5
Jul	488990	180305	179430	129255	36.87	36.69	26.43	156.1	164.0
Aug	524424	197402	192866	134156	37.64	36.78	25.58	146.0	156.1
Sep	485723	186129	179616	119978	38.32	36.98	24.70	157.5	146.0
Oct	421455	159854	158179	103422	37.93	37.53	24.54	155.4	157.5
Nov	380984	142478	143152	95354	37.40	37.57	25.03	153.9	155.4
Dec	381461	140414	153005	88042	36.81	40.11	23.08	143.6	153.9
Jan	389827	142696	151059	96072	36.61	38.75	24.64	149.7	143.6
Feb	466937	189374	171873	105690	40.56	36.81	22.63	149.9	149.7
Mar	518365	204921	186732	126712	39.53	36.02	24.44	142.6	149.9
Apr	448512	170650	163879	113983	38.05	36.53	25.41	157.5	142.6
May	543118	209181	183181	150756	38.51	33.73	27.76	169.7	157.5
Jun	534518	209598	182252	142668	39.21	34.10	26.69	156.6	169.7

The variables in the table are defined as follows.

    Total = total car sales for each month

    Low mpg car group = car sales for low mpg group (combined mpg $\leq$ 23)

    Middle mpg car group = car sales for middle mpg group (23 < combined mpg $\leq$ 26.75)

    High mpg car group = car sales for high mpg group (combined mpg > 26.75)

    Low mpg% = percentage of total car sales for low mpg group

    Middle mpg% = percentage of car sales for middle mpg group

    High mpg% = percentage of car sales for high mpg group

    Gas price of current month = average price in cents of regular gas for the month

    Gas price of previous month = average price in cents of regular gas for the month

The EESEE story "What's Driving Car Sales?" contains an electronic copy of these data, along with additional data sets and descriptions of the variables. You will need the electronic copy to answer the questions.

QUESTIONS

1. The data for exploring the relationship between car sales and gas price are average monthly gas prices, the monthly total sales data for low-, medium-, and high-mileage cars and the percentage of monthly total sales for low-, medium-, and high-mileage cars. If we want to explore whether gas price affects the sales for each of these categories of cars, would it be better to use the total monthly sales or the percentage of total monthly sales for each category of car? Explain.

2. Plot the percentage of total monthly sales for high gas mileage cars versus average monthly gas price. What trend, if any, do you observe? Are there any outliers or influential observations? What does the plot suggest about the relationship between sales of high mileage cars and gas price?

3. To determine whether the trend you observe can simply be attributed to chance, fit the least-squares regression line to the data and test whether the slope is 0. If you wish to determine whether high gas prices are associated with a higher percentage of sales of high-mileage cars, should you use a one-sided or two-sided test? Give the equation of the least-squares regression line, interpret the slope of the least-squares line, and summarize your findings.

4. Give a 95% prediction interval for the percentage of monthly total sales that is due to the sale of high-mileage cars for a month in which the price of regular gas is $1.50. How useful is this interval? Explain.

5. One might argue that there is a period of time before people react to a change in gas price and make a decision to buy a new car. Suppose this period of time is approximately one month. Repeat Questions 2 and 3, using average gas price the previous month rather than average gas price in the current month. Summarize your findings. Does current gas price or gas price the previous month appear to do a better job of explaining changes in the percentage of sales of high-mileage cars? Explain.

6. Repeat Questions 2 and 3, first exploring the relationship between monthly average gas price and percentage sales of medium-mileage cars and then exploring the relationship between monthly average gas price and percentage sales of low-mileage cars. What do you find and what do you conclude?

7. Do changes in the price of gas explain changes in car sales? Explain. Suggest any lurking variables that might be present.

For additional information and problems related to this case study, see the EESEE story "What's Driving Car Sales?"

FOLLOW-UP: Faculty Salaries (continued)

In the case study of Chapter 2, you found the least-squares regression line for predicting full professor's salaries at a given institution from the salaries of assistant professors at that institution. Test whether the slope of the least-squares regression line is 0. What does this tell you about the correlation between full and assistant professor salaries?

CHAPTER 11
MULTIPLE REGRESSION

CASE STUDY: Diamonds: How Precious Are They?

"A diamond is forever." This slogan accompanies many advertisements for diamond rings, necklaces, pendants, bracelets, and other diamond jewelry. While all diamonds are considered precious, their relative quality is determined by a combination of cut, clarity, carat weight, and color. In this case study, we explore the relationship between the

prices of 63 ladies' rings (given in Singapore dollars), the size of the diamond, measured in carat weight, and the quality of the gold in the ring. The data are given in the following table:

Observation	Size (ct)	Gold (K)	Price (SIN $$)
1	0.17	20	355
2	0.16	20	328
3	0.20	20	558
4	0.17	20	350
5	0.25	20	642
6	0.16	20	342
7	0.15	20	322
8	0.15	20	323
9	0.29	20	860
10	0.12	20	223
11	0.18	20	462
12	0.20	20	498
13	0.26	20	663
14	0.23	20	595
15	0.15	18	313
16	0.16	20	336
17	0.28	20	823
18	0.20	18	470

Observation	Size (ct)	Gold (K)	Price (SIN $$)
19	0.18	20	438
20	0.17	20	318
21	0.18	20	419
22	0.17	20	346
23	0.10	20	207
24	0.16	20	408
25	0.12	18	276
26	0.15	18	303
27	0.16	18	313
28	0.18	18	445
29	0.18	20	325
30	0.19	20	455
31	0.21	20	483
32	0.27	20	720
33	0.18	20	468
34	0.16	20	345
35	0.17	20	352
36	0.16	20	332
37	0.17	20	353
38	0.15	20	316
39	0.26	20	693
40	0.25	20	750
41	0.13	18	258
42	0.15	20	287
43	0.25	20	675
44	0.22	18	520
45	0.35	20	1086
46	0.25	20	678
47	0.15	20	315
48	0.17	20	350
49	0.32	20	918
50	0.32	20	919
51	0.15	20	298
52	0.16	20	339
53	0.16	20	338
54	0.23	20	595
55	0.21	20	589
56	0.15	18	395
57	0.23	20	553
58	0.27	20	718
59	0.18	20	443
60	0.25	20	655
61	0.33	20	945
62	0.22	20	562
63	0.17	20	345

Note 1992 conversion rate: U.S. $1.00 = SIN $1.63.

Chapter 11

The EESEE story "Diamonds: How Precious Are They?" contains an electronic copy of these data, along with additional data sets and descriptions of the variables. You will need the electronic copy to answer the questions.

QUESTIONS

1. In 1992, the conversion rate was U.S. $1.00 = SIN $1.63. Create a new variable, giving the price in U.S. dollars, based on this rate.

2. Make a scatter plot of diamond price (in U.S. dollars) versus size. Use different plotting symbols to distinguish between rings containing 18K and 20K gold. Are there any outliers or influential observations? What does your plot suggest about the relationship between the price of a lady's ring, the diamond size, and the quality of the gold in the ring? Explain.

3. Using the data, regress the price of a ring in U.S. dollars on the size of the diamond in the ring and the quality of the gold in the ring. You may first want to create an indicator variable having value 0 if the ring contains 18K gold and value 1 if the ring contains 20K gold and use this variable in your regression.

Discuss the fit of the model. Is the fit significantly better than just using the mean price of the rings as a predictor? Can either size of the diamond or quality of the gold be dropped from the model? Test the appropriate hypotheses.

4. Use the regression equation you found in Question 3 to give a 95% prediction interval for how much you would expect to pay for an 18K gold ring with a one-quarter carat diamond. How accurate is this prediction?

5. The fabled Hope Diamond weighs about 45.52 carats. Use the regression equation you found in Question 3 to give a 95% prediction interval for the price of a 20K gold ring with the Hope Diamond. Comment on your answer.

6. What would you predict the price of an 18K gold ring without a diamond (i.e., a zero-carat diamond) to be, using the regression equation in Question 3? Comment on your answer.

7. Discuss the appropriateness or inappropriateness of using linear extrapolation from this data to predict the price of a ring with very big diamonds or a ring with very tiny diamonds.

8. Repeat Question 3 including the interaction (product) between the size of the diamond and the quality of the gold in the model.

9. What other variables might affect the price of the ring? Do you think including any of these variables would dramatically improve the quality of prediction? Why?

For additional information and problems related to this case study, see the EESEE story "Diamonds: How Precious Are They?"

FOLLOW-UP: Faculty Salaries (continued)

In the case study of Chapter 2, you found the least-squares regression line for predicting full professors' salaries at a given institution from the salaries of assistant professors at that institution. You can further explore these data by using multiple regression to predict full professor salaries at a given institution from the salaries of assistant and associate professors at that institution.

FOLLOW-UP: What's Driving Car Sales? (continued)

In the case study of Chapter 10, you used simple linear regression to predict the percentage of monthly low-mileage car sales using the current month's gas price. Use multiple regression to see if this prediction is improved by adding the previous month's gas price and total monthly car sales as predictors.

CHAPTER 12

STATISTICS FOR QUALITY: CONTROL AND CAPABILITY

CASE STUDY: Control Charts for Traffic Deaths

Control charts are often employed to monitor manufacturing processes. Ccontrol charts can also be used to monitor processes in many other settings. One unusual application is traffic deaths. Yearly data on the number of motor vehicle fatalities and the number of these fatalities that were alcohol related in the province of Manitoba, Canada, are given in the following table.

Year	Number of motor vehicle fatalities	Number. of alcohol-related fatalities	Year	Number. of motor vehicle fatalities	Number. of alcohol-related fatalities
1973	249	123	1982	160	72
1974	220	98	1983	155	62
1975	226	121	1984	143	56
1976	226	122	1985	144	64
1977	185	90	1986	184	99
1978	208	84	1987	186	80
1979	179	115	1988	156	74
1980	184	82	1989	173	74
1981	224	120	1990*	38	16

*Data through June 30 only

Spiring (1994) (see EESEE for the exact reference) uses control charts to explore alcohol related traffic deaths before and after a tough drinking and driving law was enacted in Manitoba in 1989. In this case study, we will also explore this use of control charts.

The EESEE story "Control Charts for Traffic Deaths" contains an electronic copy of these data, along with additional data sets and descriptions of the variables. You will need the electronic copy of the data to answer the questions.

QUESTIONS

1. Presumably, one objective of the drinking and driving law was to reduce alcohol-related fatalities. To explore the effect of the law, one could look at the annual number of alcohol-related motor vehicle fatalities or at the annual proportion of all motor vehicle fatalities that are alcohol related. Discuss the advantages and disadvantages of these two measures.

2. Assuming that the law was not passed at the beginning of 1989, are the 1989 data useful for assessing the effect of the law? Why or why not?

3. For each of the years 1973 to 1990, compute the proportion of motor vehicle fatalities that are alcohol related.

4. Think of the data in the table as coming from a process. We might view the data (in particular, the proportion of motor vehicle fatalities that are alcohol related) before 1989 as representing this process prior to passage of the drinking and driving law. If the law significantly reduces alcohol-related fatalities, we would expect the proportion of motor vehicle fatalities that are alcohol related after 1989 to display a downward shift in the mean of this process. A p-chart, constructed from the data before 1989, could be used to see if this is the case. What would the center line be for such a control chart? For determining whether the proportion in 1990 (the only data in the table that can be used to assess the effect of the law) is out of control, what control limits would you use? Does the proportion in 1990 appear to be out of control?

5. Plot the proportion of motor vehicle fatalities that are alcohol related against time for the period 1973 to 1988. What, if any, patterns do you notice? Is it reasonable to assume that the proportion of motor vehicle fatalities that are alcohol related was stable (relatively constant) over this period? Based on the plot, would a downward shift over the next several years in the proportion of motor vehicle fatalities that are alcohol related be good evidence that the law is effective? Discuss.

6. Describe how you might construct a p-chart to monitor future observations to assess the effect of the law over time. What patterns would lead you to believe the law was effective?

7. After comparing the data from November 1989 to June 1990 to the data from November 1988 to June 1989, the government of Manitoba announced that after drunk driving legislation was passed in November 1989 the number of alcohol-related fatalities had been almost cut in half. Our data here are not classified according to those time periods but instead are classified yearly. Are your conclusions, based on the previous problems, similar to those of the government report? Can you suggest a reason why comparing two periods of November to June with each other could yield different results from comparing the period January to December in previous years to the period January to June in 1990?

For additional information and problems related to this case study, see the EESEE story "Control Charts for Traffic Deaths."

CHAPTER 13
TIME SERIES FORECASTING

CASE STUDY: The Changing Face of Agriculture

Agricultural crop production is one of the most important sectors of the U.S. economy. For example, the two most valuable crops, corn and soybeans, have a production value of about $19 billion and $13 billion per year, respectively. Mimicking other sectors of the economy, advancing technology has brought sweeping advances in crop productivity, while at the same time increasing concentration has replaced small family-owned farms with large corporations. Along with these general trends, crop production is also affected by the seasonal nature of farming and the more random fluctuations in global markets and in the weather. The Economic Research Service (ERS) of the U.S. Department of Agriculture is charged with keeping track of this volatile part of the economy and forecasting future production and prices. This information is needed for a wide variety of purposes, from making decisions about loans to farmers to preparing the budget for the food stamp program, understanding how energy needs might be met by ethanol production, and negotiating foreign aid and trade agreements. A great deal of effort goes into this time series forecasting and major reports on the outlook for crops are posted monthly at the ERS website at www.ers.usda.gov.

Ohio is one of the key crop-producing states with a Department of Agriculture that supplies data to the ERS. The state of Ohio mirrors the national trends in increasing crop yields despite a shrinking number of farms. Also, as in the nation as a whole, corn and soybeans are the top money-producing crops in Ohio. The following table provides data from the Ohio Department of Agriculture on the state's corn and soybean crops between 1942 and 1994.

Chapter 13

Year	Number of farms (thousands)	Corn acreage (thousands of acres)	Corn yield (bushels/acre)	Soybean acreage (thousands of acres)	Soybean yield (bushels/acre)
1942	236	3249	56	1051	22
1943	230	3444	51	1213	21
1944	225	3651	40	1243	18
1945	223	3468	50	1077	18
1946	219	3641	49	903	18
1947	216	3386	41	950	18
1948	214	3691	58	908	20
1949	211	3617	54	875	24
1950	208	3364	52	1090	22
1951	200	3532	48	1124	19
1952	192	3567	53	940	22
1953	186	3531	55	1007	20
1954	180	3708	61	1122	25
1955	174	3708	59	1193	24
1956	168	3523	60	1301	24
1957	162	3206	54	1404	23
1958	158	3206	60	1429	26
1959	154	3687	62	1455	25
1960	149	3576	68	1499	24
1961	144	2718	74	1722	28
1962	139	2960	76	1756	25
1963	135	3200	78	1738	24
1964	131	3328	65	1825	22
1965	129	3295	74	2044	24
1966	126	3328	84	2105	28
1967	124	3516	79	2231	22
1968	122	3096	86	2325	30
1969	120	3013	87	2475	30
1970	118	3249	79	2550	28

1971	116	3776	91	2634	30
1972	114	3286	92	3010	26
1973	113	3270	80	3590	26
1974	112	3820	75	3140	26
1975	101	3575	93	3100	33
1976	99	4080	103	2880	33
1977	97	3870	105	3480	35
1978	96	3850	105	3870	33
1979	96	3830	115	4080	36
1980	95	4120	113	3760	36
1981	94	4040	96	3450	28
1982	93	4255	114	3700	36
1983	92	3060	80	3280	32
1984	90	4130	118	3770	36
1985	89	4240	127	3870	42
1986	88	3880	128	3620	40
1987	84	3260	120	3900	37
1988	85	3250	85	3700	27
1989	86	3140	118	3980	32
1990	84	3450	121	3480	39
1991	80	3400	96	3770	36
1992	78	3550	143	3680	40
1993	76	3280	110	4110	38
1994	75	3500	139	3990	44

The EESEE story "Historical Farm Data" contains an electronic copy of these data, along with additional data sets and descriptions of the variables. You will need the electronic copy to answer the questions.

QUESTIONS

1. Make time series plots of the soybean and corn yield data and describe the patterns you see. Which crop experienced greater increases in productivity over the decades covered by the data? Explain.

2. Make a time series plot of the number of farms and of the number of acres planted in corn.

a) Describe the patterns you see in the plots. For which variable would forecasting future values be more difficult? Explain.

b) In 1974, the U.S. Department of Agriculture made a subtle change in the definition of a farm. Was the new definition more conservative or was the definition expanded? Explain how you can answer this question from your time series plot.

c) Are Ohio's corn farms getting larger? Or is it impossible to tell from these graphs? Explain.

3. Use regression to fit the overall trend in the number of bushels per acre produced by Ohio's soybean farmers. How do you interpret the slope in the context of soybean yields?

4. In time series forecasting, it is often suggested to start the time variable at one (for example by subtracting 1941 from all the years in this data). Fit the regression of soybean yields on time with the time variable starting at one. How does the intercept in this regression compare to the one in Question 3 where the year was entered directly as the explanatory variable? In which case does the intercept have a more natural interpretation? Explain.

5. Make a lagged residual plot after adjusting for the overall trend in the soybean yield data. Is there a strong pattern of autocorrelation? Explain what this mean in terms of forecasting future soybean yields.

6. Make a time series plot of how many thousands of acres of Ohio land are planted in soybeans and describe the pattern you see. What features of the time series plot make forecasting difficult? Make a lagged residual plot after adjusting for the overall trend in the number of acres of soybeans. Is there a strong pattern of autocorrelation? Explain what this mean in terms of forecasting future years.

7. In the year 2000, Ohio harvested 3,480,000 acres of corn with an average yield of 147 bushels per acre. Based on forecasts that might have been made back in 1994, was the year 2000 a good year or a bad year for Ohio's corn producers? Explain.

For additional information and problems related to this case study, see the EESEE story "Historical Farm Data."

FOLLOW-UP: WHAT'S DRIVING CAR SALES (continued)

In the case study in of Chapter 10, what was the average monthly number of cars sold between January 2000 and June 2001? Based on the overall trend in car sales during this period, what would you predict for the total sales of cars in July 2001? Car sales vary with the season as new models are introduced on a regular schedule. Adjusting for this seasonality, what would you predict for the total sales of cars in July 2001? Look up the actual total car sales for July 2001. Were your predictions closer to the truth than just predicting the overall average?

CHAPTER 14
ONE-WAY ANALYSIS OF VARIANCE

CASE STUDY: Fighting for Shelf Space

Americans consume an average of about 13 pounds of ready-to-eat breakfast cereal per person per year, so it is no surprise that most large supermarkets devote an entire aisle just to cereals. With over 400 different brands of  cereal available to put on the shelves, however, there is fierce competition for prime shelf space. Major manufacturers are eager to negotiate with retailers for top shelf placements by offering discounts, paying for advertising in local store flyers, paying a special "slotting fee" to get new products on the shelf, and even by providing free research into consumer buying habits. Retailers and manufacturers often find common ground. The manufacturers realize that products in the prime locations sell best, and at the same time retailers want to reserve the best shelf locations for the hottest-selling items. Bottom shelves are shunned. The middle shelves, at the eye level of small children, are thought to be the best place to position kids' cereals. Top shelves at the eye level of adult shoppers are usually the most sought-after territory.

How does this competition for a share of the $10 billion breakfast cereal market distribute cereals across the shelves? Do store brands and cereals made by small producers tend to fall on the bottom shelves? Does the nutritional value of the cereals vary from shelf to shelf based on the targeted consumers? We can examine this last question using the nutritional information required by the FDA that is placed on the sides of cereal boxes. The following table provides the number of grams of sugar per serving for 76 different breakfast cereals, along with the supermarket's shelf placement.

One-Way Analysis of Variance

Cereal	Sugar (gr/serv)	Shelf	Cereal	Sugar (gr/serv)	Shelf
100% Bran	6	top	Just Right Crunchy Nuggets	6	top
100% Natural Bran	8	top	Just Right Fruit & Nut	9	top
All-Bran	5	top	Kix	3	middle
All-Bran with Extra Fiber	0	top	Life	6	middle
Almond Delight	8	top	Lucky Charms	12	middle
Apple Cinnamon Cheerios	10	bottom	Maypo	3	middle
Apple Jacks	14	middle	Mueslix Raisins, Dates, & Almonds	11	top
Basic 4	8	top	Mueslix Raisins, Peaches, & Pecans	11	top
Bran Chex	6	bottom	Mueslix Crispy Blend	13	top
Bran Flakes	5	top	Multi-Grain Cheerios	6	bottom
Cap 'n' Crunch	12	middle	Nut & Honey Crunch	9	middle
Cheerios	1	bottom	Nutri-Grain Almond-Raisin	7	top
Cinnamon Toast Crunch	9	middle	Nutri-grain Wheat	2	top
Clusters	7	top	Oatmeal Raisin Crisp	10	top
Cocoa Puffs	13	middle	Post Nat. Raisin Bran	14	top
Corn Chex	3	bottom	Product 19	3	top
Corn Flakes	2	bottom	Puffed Rice	0	top
Corn Pops	12	middle	Puffed Wheat	0	top
Count Chocula	13	middle	Quaker Oat Squares	6	top
Cracklin' Oat Bran	7	top	Raisin Bran	12	middle
Cream of Wheat (Quick)	0	middle	Raisin Nut Bran	8	top
Crispix	3	top	Raisin Squares	6	top
Crispy Wheat & Raisins	10	top	Rice Chex	2	bottom
Double Chex	5	top	Rice Krispies	3	bottom
Froot Loops	13	middle	Shredded Wheat	0	bottom
Frosted Flakes	11	bottom	Shredded Wheat 'n' Bran	0	bottom
Frosted Mini-Wheats	7	middle	Shredded Wheat spoon size	0	bottom
Fruit & Fibre	10	top	Smacks	15	middle
Fruitful Bran	12	top	Special K	3	bottom
Fruity Pebbles	12	middle	Strawberry Fruit Wheats	5	middle
Golden Crisp	15	bottom	Total Corn Flakes	3	top
Golden Grahams	9	middle	Total Raisin Bran	14	top
Grape Nuts Flakes	5	top	Total Whole Grain	3	top
Grape-Nuts	3	top	Triples	3	top
Great Grains Pecan	4	top	Trix	12	middle
Honey Graham Ohs	11	middle	Wheat Chex	3	bottom
Honey Nut Cheerios	10	bottom	Wheaties	3	bottom
Honey-comb	11	bottom	Wheaties Honey Gold	8	bottom

Chapter 14

The EESEE story "Nutrition and Breakfast Cereals" contains an electronic copy of these data, along with additional data sets and descriptions of the variables. You will need the electronic copy to answer the questions.

QUESTIONS

1. Make side-by-side boxplots of the sugar content of the cereals by shelf position. Describe what this graph tells you about how sugar content varies with shelf position. Does the pattern in the graph make sense given the nature of shelf placements? Explain.

2. Estimate the mean grams of sugar per serving for each of the three shelf positions. Does it appear reasonable to assume a common variance in sugar content from shelf to shelf? If no, explain why not. If yes, calculate an estimate of this common variance.

3. Introduce appropriate notation and write out a statistical model that approximately describes the sugar content data in the table. Use your notation to specify the null and alternative hypotheses that would be tested by someone investigating whether sugar content varies with shelf position.

4. Using statistical software of your choice, produce the ANOVA table to examine the within-shelf and between-shelf variability in cereal sugar content.

a) Explain how each column of the table is related to the other columns (for example, explain how the mean squares column is related to the sum of squares and degrees of freedom columns).

b) Give the P-value for testing the null hypothesis from Question 3 and provide a clear interpretation of what this says about whether the mean sugar content of cereals changes with shelf position.

5. Evaluate the assumptions underlying the use of the analysis of variance technique for these data. Does a normal quantile plot show that the residuals are approximately normal? Is the rule of thumb for equality of population variances satisfied? Is it reasonable to view the data as a random sample from an overall shelf placement process? Explain.

6. High-sugar cereals are thought to appeal to children, and cereal manufacturers often try to position children's cereals on the middle shelf. Thus, it might be expected that the mean sugar content of cereals on the middle shelf would be higher than the average of the mean sugar content of the top and bottom shelves. Write an expression for the contrast that would be used to investigate this assumption. Find a 95% confidence interval for this contrast and summarize your results as you might in a report to a supermarket manager.

For additional information and problems related to this case study, see the EESEE story "Nutrition and Breakfast Cereals."

FOLLOW-UP: DRIVE-THRU COMPETITION (continued)

In the case study of Chapter 8, do the top five national fast-food restaurant chains take an equal amount of time to serve drive-thru customers? Explain how ANOVA could be used to investigate this question using data from the g3 Mystery Shopping study. Is there sufficient information in the data set given with the Chapter 8 case study to perform this analysis? If no, explain what information is needed. If yes, carry out the ANOVA and interpret your results.

CHAPTER 15

TWO-WAY ANALYSIS OF VARIANCE

Currently we do not have good business examples of two-way analysis in EESEE. Thus, we have not included a case study for this chapter. We anticipate adding appropriate examples to EESEE soon and a case study involving two-way analysis of variance to this book in a future edition. In the meanwhile, the EESEE story "Stepping Up Your Heart Rate" can be used to explore two-way analysis of variance.

CHAPTER 16

NONPARAMETRIC TESTS

CASE STUDY: Floral Scents and Learning

Do certain odors affect learning? Can pleasant aromas help a student concentrate? Hirsch and Johnston from the Smell and Taste Treatment and Research Foundation, Chicago, conducted a study to determine if the presence of a floral scent can improve a person's learning ability in certain situations. They found that most people showed greater improvement on their times through a pencil-and-paper maze when they were exposed to a floral scent than when they weren't.

The study was conducted as follows. Twenty-two people worked through a set of two pencil-and-paper mazes six times, three times on one maze while wearing a floral-scented mask and three times on a second maze wearing an unscented mask. The three trials for each mask closely followed one another. Individuals were randomly assigned to wear the floral mask on either their first three tries or their last three tries. Participants put on their masks one minute before starting the first trial in each group to minimize any distracting effect. Subjects recorded whether they found the scent inherently positive or inherently negative or if they were indifferent to it. Researchers measured the length of time it took subjects to complete each of the six trials. To ensure that the subjects could actually smell the floral scent, they were each given two tests measuring olfactory ability. The scores of one person were removed from the data set because he failed both smell tests.

The measurements on the remaining twenty-one subjects are given in the following table.

ID	Gender	Age	Smoker	Opinion	Order	U-trial 1	U-trial 2	U-trial 3	S-trial 1	S-trial 2	S-trial 3
1	M	23	N	pos	1	38.4	27.7	25.7	53.1	30.6	30.2
2	F	43	Y	neg	2	46.2	57.2	41.9	54.7	43.3	56.7
3	M	43	N	pos	1	72.5	57.9	51.9	74.2	53.4	42.4
4	M	32	N	neg	2	38.0	38.0	32.2	49.6	37.4	34.4
5	M	15	N	neg	1	82.8	57.9	64.7	53.6	48.6	44.8
6	F	37	Y	pos	2	33.9	32.0	31.4	51.3	35.5	42.9
7	F	26	N	pos	1	50.4	40.6	40.1	44.1	46.9	42.7
8	F	35	N	pos	2	35.0	33.1	43.2	34.0	26.4	24.8
9	M	26	N	pos	1	32.8	26.8	33.9	34.5	25.1	25.1
10	F	31	N	indiff	2	60.1	53.2	40.4	59.1	87.1	59.2
11	F	35	Y	pos	1	75.1	63.1	58.0	67.3	43.8	42.2
12	F	55	Y	indiff	2	57.6	57.7	61.5	75.5	126.6	48.4
13	F	25	Y	pos	1	55.5	63.3	44.6	41.1	41.8	32.0
14	M	39	Y	indiff	2	49.5	45.8	35.3	52.2	53.8	48.1
15	M	25	N	indiff	1	40.9	35.7	37.2	28.3	26.0	33.7
16	M	26	N	pos	2	44.3	46.8	39.4	74.9	45.3	42.6
17	M	33	Y	neg	1	93.8	91.9	77.4	77.5	55.8	54.9
18	M	62	N	neg	2	47.9	59.9	52.8	50.9	58.6	64.5
19	F	54	Y	pos	1	75.2	54.1	63.6	70.1	44.0	43.1
20	F	38	N	neg	2	46.2	39.3	56.6	60.3	47.8	52.8
21	M	65	N	neg	1	56.3	45.8	58.9	59.9	36.8	44.3

ID = Identification number

Gender = Subject's gender. "F" = female, "M" = male

Age = Subject's age in years

Smoker = Y if subject smoked, N if subject did not smoke

Opinion = "pos" if subject found the odor inherently positive, "indiff" if subject was indifferent, "neg" if subject found the odor inherently negative

Order = 1: the subject wore the unscented mask on the first three trials. 2: the subject wore the unscented mask on the last three trials.

U-trial 1, U-trial 2, U-trial 3, S-trial 1, S-trial 2, S-trial 3 = length of time required for a trial. U-trial stands for trials with no scent present (unscented trials), S-trial for trials in which the scent was present.

The EESEE story "Floral Scents and Learning" contains an electronic copy of these data along with additional data sets and descriptions of the variables.

The two mazes follow. Each may be completed by connecting the 25 numbered circles in order.

Nonparametric Tests 83

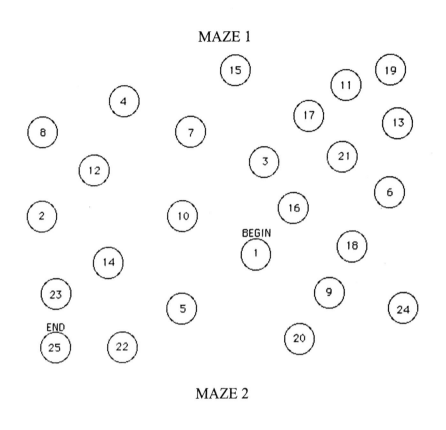

QUESTIONS

1. The question of interest to the researchers was whether the scent enhances a person's ability to learn. Learning could be measured by computing the percent decrease in the time to complete a maze from trial 1 to trial 3. For example, the percent decrease in time to complete the maze while wearing the unscented mask is

$$\text{unscented percent decrease} = \frac{\text{U-trial 1} - \text{U-trial 3}}{\text{U-trial 1}} \times 100\%$$

What are the advantages of using percentage decrease as a measure of learning rather than the net decrease U-trial1 − U-trial3?

2. Compute these percent decreases for both the unscented and scented trials for each subject. Prepare a graphical display to see if there is any evidence that there is a greater improvement in learning (greater percent decrease in the time to complete the maze) when wearing the scented mask.

3. Use graphic displays to explore whether any of the categorical explanatory variables (gender, smoker, opinion, order) affect the percentage decrease from trial 1 to trial 3 for the unscented and scented groups. For which of these variables does there appear to be a significant separation between the scented and unscented groups? Explain.

4. Prepare a plot to see if age has an effect on the magnitude of the difference between the two groups. Does age appear to be useful as a predictor? Explain.

5. Perform appropriate nonparametric tests to see if there is a significant difference between the medians for the two percentage decrease variables for those subjects classified according to each of the categorical variable that you identified in Question 3. What do you conclude?

6. The next time you study, will you do so in the presence of a pleasant floral scent? Why or why not?

For additional information and problems related to this case study, see the EESEE story "Floral Scents and Learning."

CHAPTER 17

LOGISTIC REGRESSION

Currently we do not have good business examples of logistic regression in EESEE. Thus, we have not included a case study for this chapter. We anticipate adding appropriate examples to EESEE soon and a case study involving logistic regression to this book in a future edition. In the meanwhile, any story involving two by two tables can be used to explore logistic regression. The EESEE story "Trilobite Bites" is one such story.

CHAPTER 18

BOOTSTRAP METHODS OR PERMUTATION TESTS

CASE STUDY: Is Friday the 13th Unhealthy?

Is Friday the 13th an unusually unlucky day? How do superstitions, such as the belief that Friday the 13th is unlucky, affect people's behavior? These questions were addressed in a study that examined the relationship between health, behavior, and superstition in the United Kingdom. Shopping and traffic patterns were analyzed, as were the numbers and types of accidents that occurred on Friday the 13th.

The study focused on two questions:

1: How do superstitions regarding Friday the 13th affect human behavior?

2: Is Friday the 13th unluckier than other Fridays?

We will examine only the first question in this case study. We will do so by comparing the number of shoppers on all Friday the 13th's between July of 1990 and November of 1992 to the number on the Friday of the week before in nine different supermarkets in southeast England. If shopping patterns are different on Friday the 13th than on other Fridays, it would be good for store owners to know it.

The following table provides information collected for the study.

Date	Location	Number of shoppers Fri. 6th	Number of shoppers Fri. 13th
1990, July	Epsom	4942	4882
1991, September	Epsom	4895	4736
1991, December	Epsom	4805	4784
1992, March	Epsom	4570	4603
1992, November	Epsom	4506	4629
1990, July	Guildford	6754	6998
1991, September	Guildford	6704	6707
1991, December	Guildford	5871	5662
1992, March	Guildford	6026	6162
1992, November	Guildford	5676	5665
1990, July	Dorking	3685	3848
1991, September	Dorking	3799	3680
1991, December	Dorking	3563	3554
1992, March	Dorking	3673	3676
1992, November	Dorking	3558	3613
1990, July	Chichester	5751	5993
1991, September	Chichester	5367	5320
1991, December	Chichester	4949	4960
1992, March	Chichester	5298	5467
1992, November	Chichester	5199	5092
1990, July	Horsham	4141	4389
1991, September	Horsham	3674	3660
1991, December	Horsham	3707	3822
1992, March	Horsham	3633	3730
1992, November	Horsham	3688	3615
1990, July	East Grinstead	4266	4532
1991, September	East Grinstead	3954	3964
1991, December	East Grinstead	4028	3926
1992, March	East Grinstead	3689	3692
1992, November	East Grinstead	3920	3853
1990, July	Lewisham	7138	6836
1991, September	Lewisham	6568	6363
1991, December	Lewisham	6514	6555
1992, March	Lewisham	6115	6412
1992, November	Lewisham	5325	6099
1990, July	Nine Elms	6502	6648
1991, September	Nine Elms	6416	6398
1991, December	Nine Elms	6422	6503
1992, March	Nine Elms	6748	6716
1992, November	Nine Elms	7023	7057
1990, July	Crystal Palace	4083	4277
1991, September	Crystal Palace	4107	4334
1991, December	Crystal Palace	4168	4050
1992, March	Crystal Palace	4174	4198
1992, November	Crystal Palace	4079	4105

The EESEE story "Is Friday the 13th Unhealthy?" contains an electronic copy of these data, along with additional data sets and descriptions of the variables. You will need the electronic copy to answer the questions.

QUESTIONS

1. These data are paired. For each case given, compute the difference between the number of shoppers on Friday the 6th and Friday the 13th; i.e., compute the quantity

number of shoppers on Friday the 6th − number of shoppers on Friday the 13th.

Make a histogram of these differences. Describe the shape of the distribution. Are there any unusual features? Explain.

2. Do people change their behavior on Friday the 13th? We can use the data given in this case study to investigate this question. Formulate what you consider to be appropriate null and alternative hypotheses about the mean difference in the number of shoppers for testing whether Friday the 13th affects people's shopping behavior.

3. Does it make sense to do a paired t test to test your hypotheses? If yes, carry out the test and report your conclusions. If not, explain why not.

4. Carry out a permutation test for matched pairs to test your hypotheses. What do you conclude?

5. Many people are paid biweekly, i.e., on the second and fourth Fridays of the month. If this is the case for a large number of people in England, we might expect to see the effect in the data. Formulate what you consider to be appropriate null and alternative hypotheses regarding the mean difference in the number of shoppers for testing whether being paid biweekly affects shopping behavior.

6. Carry out a permutation test for matched pairs to test your hypotheses. What do you conclude?

For additional information and problems related to this case study, see the EESEE story "Is Friday the 13th Unhealty?"